Student's Book About Creation

by

E. Norbert Smith, Ph.D.

Illustrated by Nathalie Kelley

Student's Book About Creation
2014 Copyright by E. Norbert Smith, Ph.D.

Printed in the United States if America

Smith, E. Norbert, Ph.D.   1941

ISBN-13 978-1502593078

ISBN-10 1502593076

Tonya Holmes Shook: Publishing Coordinator

Table of Contents

Dedication

Then God said, "Let the land produce vegetation: seed-bearing plants and trees on the land that bear fruit with seed in it, according to their various kinds." And it was so. (Genesis 1:11)

A beautiful enchanted forest in Oregon.
Photo by Norbert Smith

God created a beautiful world for us to enjoy and filled it with many fascinating plants and animals. This book is dedicated to children and adults that love God's Creation and want to learn more about its wonders. Our hope and is this book will motivate many young people to become interested in science so they can learn more about God's wonderful Creation.

Consider how the lilies grow. They do not labor or spin. Yet I tell you, not even Solomon in all his splendor was dressed like one of these. (Luke 12:27)

Water Lily blossom
Photo by Ginger Read

Friends that helped

Plans fail for lack of counsel, but with many advisers they succeed. (Proverbs 15:22)

We all need the help, advice and encouragement from others. Several people have helped make this project possible. Tonya Shook has encouraged, motivated and helped me with the publication process and I could not have done it without her. My friend and neighbor Ginger Read is a retired school principal. She has often encouraged and sometimes nagged me to complete this project. She also provided the photo of the beautiful lily in the Dedication section.

Sean and Julie Williams and Julie's parents, Gene and Oma Hicks have been my longtime friends during good times and bad. The beautiful cover photo of the ornate box turtle was taken by Sean Williams. I appreciate the help of each of you and many others that have inspired me to study God's Creation over my lifetime. Life is good and friends add zest.

God protects us

He will cover you with his feathers. He will shelter you with his wings. His faithful promises are your armor and protection. There is absolutely nothing to fear about tomorrow; for God is already there.
Psalms 91:4

**Mother bird feeding and protecting her babies.
God's lesson through nature.**

Ask the animals

But ask the animals, and they will teach you, or the birds of the air, and they will tell you; or speak to the earth, and it will teach you, or let the fish of the sea inform you. Job 12:7-8

God gave us inquiring minds with the ability to think and learn. We all have many questions about the world around us. Some of the questions are easy and some are hard. Here are a few hard questions. Where did I come from? Why am I here? What is my purpose? God gave us the ability to find the answers to these important questions.

You can learn many things from our mother and father. You can learn from your teachers at school. As the Bible says, we can also learn from the world around us. The Bible tells us to ask questions of living things and of the earth. One of your authors, Norbert Smith has spent much of his life asking questions animals. They have told him many things.

We can also learn from reading books. People wrote all the books except the Bible. God wrote the Bible. He used people to write down the words, but He told them what to write. Because God wrote the Bible, it can be trusted. What it says is true. The Bible has the answers to many of life's difficult questions.

How it all began

In the beginning God created the heavens and the earth. (Gen 1:1. NIV)

The Bible begins with these words, but some people today do not believe this important teaching. They think the world has always been here and that plants, animals and even people came along by accident. Those people are wrong. All living things are very complicated and could not have been made by accident. Neither could they have made themselves. They had to have a Designer and that Designer is the God and Creator described in the Bible.

Here is an example that illustrates this important point. It was first used over 200 years ago by William Paley from England. If someone is walking outdoors and finds a pocket watch it is obvious that someone must have dropped the watch. A watch has many tiny parts and could not make itself. So it is with living things. They are more complicated than a watch and must have had a designer. That Designer is God. He created all things and you can know Him too.

Scientists can be wrong and the Bible warns us about this. ***O Timothy, keep that which is committed to thy trust, avoiding profane and vain babblings, and oppositions of science falsely so called: which some professing have erred concerning the faith.*** (1 Timothy 6:20-21).

The Bible teaches that God created the world in six days. Remember, we can trust the Bible. Let's look and see what the Bible says about how our world began.

Part One: The First Week

In the beginning God…

Contrary to what some scientists believe, the universe has not always been in existence. There WAS a beginning. The Bible starts with the important words above which sets the stage for all that follows. Read and learn the truth about the world God made for all of us.

The First Monday

In the beginning God created the heavens and the earth. Now the earth was formless and empty, darkness was over the surface of the deep, and the Spirit of God was hovering over the waters. And God said, "Let there be light," and there was light. God saw that the light was good, and he separated the light from the darkness. God called the light "day," and the darkness he called "night." And there was evening, and there was morning-- the first day. (Genesis 1:1-5)

From NASA, the National Aeronautics and Space Administration.

This is how the world began. It did NOT just happen. There were no animals. There were no people. There were no trees or flowers. There were no mountains or rivers. There was no light. There was only God. He alone was there in the beginning and knows how our world was made. We can trust God. Man can only guess, but God was there when it happened.

Spiral nebula, 3mpub.com/nadel from NASA.

Only God could create the heavens and the earth by just His word. He created light on the first day of creation. Light is important. Plants must have light to grow. Plants provide food and make the oxygen we need to breathe. Without light there can be no life. Jesus understood the importance of light. He said, *"I am the light of the world."* (John 8:12) Just like the world needs light, so we need Jesus as our Savior.

Evening and morning marks each of the days of creation. It is the rotation of the earth that determines the length of days. Some people think the days were long periods of time, but they are wrong. These days were regular days like we have now. The proof that the days of creation were regular days is found in Exodus. That is the second book in the Bible. The Ten Commandments are found in Exodus. One of the Ten Commandments says that we are not supposed to work on the Sabbath day or Sunday.

For in six days the LORD made the heavens and the earth, the sea, and all that is in them, but he rested on the seventh day. Therefore the LORD blessed the Sabbath day and made it holy. (Exodus 20:11)

Since Sunday is a normal 24-hour day so are the other six days.

The First Tuesday

And God said, *"Let there be an expanse between the waters to separate water from water." So God made the expanse and separated the water under the expanse from the water above it. And it was so. God called the expanse "sky. And there was evening, and there was morning-- the second day. (Genesis 1:6-8)*

There are many Bible verses that expand with this important event. Here is my favorite.

Praise him, you highest heavens and you waters above the skies. Let them praise the name of the LORD, for he commanded and they were created. (Psalms 148:4-5)

As we have seen, God created the world out of nothing. Fish and birds were made from water. Animals and people were made from the earth, but God made the world out of nothing, but his spoken Word. We worship an all-powerful God and there is nothing He cannot do. Let us praise and honor him with all our strength.

Water is important for all living things just like the light that God created on the first day. We can live a month without food, but will die in only 4 days without water to drink. The Bible confirms the importance of water. The prophet Isaiah said:

The LORD will guide you always; he will satisfy your needs in a sun-scorched land and will strengthen your frame. You will be like a well-watered garden, like a spring whose waters never fail. (Isaiah 58:11)

Jesus compared belief in Him to a stream of water.

Whoever believes in me, as the Scripture has said, streams of living water will flow from within him." (John 7:38)

The First Wednesday

A peaceful lake in Colorado.
Photo by Norbert Smith

And God said, "Let the water under the sky be gathered to one place, and let dry ground appear." And it was so. God called the dry ground "land," and the gathered waters he called "seas." And God saw that it was good. Then God said, "Let the land produce vegetation: seed-bearing plants and trees on the land that bear fruit with seed in it, according to their various kinds." And it was so. The land produced vegetation: plants bearing seed according to their kinds and trees bearing fruit with seed in it according to their kinds.

And God saw that it was good. And there was evening, and there was morning-- the third day. (Genesis 1:9-13)

The excitement is building. Light and water were created on the first and second day of Creation. On the third day the land and seas were separated. Dry land was made and provided a place many living things to live. On the third day plants were created. Plants make oxygen and all animals must have oxygen to survive.

Plants are important for another reason. They provide food for animals and people. Animals eat plants or they eat animals that eat plants. We cannot live without light and water. We must also have food. God is preparing the earth so animals and people will have a place to live and food to eat. Plants also add beauty to our world.

Morning Glory.

There is another important thing mentioned on the third day of Creation. The plants were commanded to reproduce and make more plants like themselves. Just as dogs produce dogs and cats produce cats, so plants produce other plants. We understand how this happens. It is because of genetics. Genetic works because of the genetic code and is a written language. That language uses DNA as the words. God is not only our Creator. He is also the Author of all of living things. A good cook writes a recipe for what he or she makes. God wrote the recipe for all the plants and animals that He created in the form of DNA.

Tiger Lily.

Flowers appear on the earth; the season of singing has come, the cooing of doves is heard in our land.
(Song of Solomon 2:12)

The First Thursday

Full moon rising.
Photo by Tim McCord

And God said, "Let there be lights in the expanse of the sky to separate the day from the night, and let them serve as signs to mark seasons and days and years, and let them be lights in the expanse of the sky to give light on the earth." And it was so. God made two great lights-- the greater light to govern the day and the lesser light to govern the night. He also made the stars. God set them in the expanse of the sky to give light on the earth, to govern the day and the night, and to separate light from darkness. And God saw that it was good.

And there was evening, and there was morning-- the fourth day. (Genesis 1:14-19)

God made the sun, moon and stars on the fourth day of Creation. The sun gives us light during the day. The moon and stars provide light at night. Light is important. Without light we could not see. Plants must have light to live. People and animals need plants for food and oxygen to breath. There is another important thing these things do. They also tell us the seasons of the year. This is important so we know when to plant seeds and harvest crops.

Neil Armstrong

Buzz Aldrin on the moon

For thousands of years people looked at the moon and wondered what it would be like to go there. On July 20, 1969 we found out. Two astronauts, Neil Armstrong and Buzz Aldrin, landed on the moon and spent over two hours walking around and exploring it.

Astronauts have made six trips to the moon from 1969 to 1972. They have brought back nearly 50 pounds soil and rock samples from for scientists to study. We must remember that all the lights of heaven are the work of God's hands.   He made them. Without the light from the sun, life would not be possible.

My help comes from the LORD, the Maker of heaven and earth. (Psalms 121:2)

The First Friday

And God said, "Let the water teem with living creatures, and let birds fly above the earth across the expanse of the sky." So God created the great creatures of the sea and every living and moving thing with which the water teems, according to their kinds, and every winged bird according to its kind. And God saw that it was good. God blessed them and said, "Be fruitful and increase in number and fill the water in the seas, and let the birds increase on the earth." And there was evening, and there was morning-- the fifth day. (Genesis 1:20-23)

Birds and all creatures that live in water were created on this day. Birds are important for several reasons. Their flight is amazing and always a joy to watch. They helped motivate people to invent airplanes so we could also fly. They add beauty and their singing lifts our spirits. Perhaps most importantly they eat many insect pests. Many birds such as doves, turkeys, chickens and waterfowl are good to eat and provide food for us.

For years man wanted to fly like a bird. After many unsuccessful attempts man was finally able to fly and birds helped provide the motivation. In 1903 Orville Wright was the first human to fly in an airplane. The flight occurred in North Carolina and only lasted 12 seconds, but that is when modern aviation was born.

Many birds do something that people have found interesting for over 3,000 years. They migrate with the changing seasons. As the days grow short many birds leave their summer homes and fly south to where the weather is warmer. The Arctic Tern is a shore bird that lives around water. It holds the record for the longest yearly migration. Each year terns travel from the Arctic to Antarctica and back. This is a distance of 12,000 miles.

Fish are also important and were made on the first Friday. They live in the oceans as well as lakes, ponds, rivers and small streams. Many fish are also beautiful to watch and provide food for various birds and other animals. Human beings enjoy eating many kinds of fish. A number of people have aquariums and are fascinated watching brightly colored fish or just watching other animals that live in water.

Fishing also provides people with an opportunity to spend outside in God's Creation. My Grandfather and Grandmother enjoyed fishing and as a young boy, I often went with them. Fishing also provided good food for us to eat. Not only does fish taste good, but it also provides many important nutrients we need to stay healthy. Fish contains less fat and oils than other kinds of meat and is very good for you to eat. There is a huge fishing industry along our coasts that provides food for many people living far from the oceans. Many years ago I owned part of a fishing boat in Maine and enjoyed many hours fishing in the Atlantic Ocean. Lobster and crabs are still two of my favorite foods. I have also spent some time digging for claims when the tide was out. Because I live in Oklahoma, I no longer see the ocean and miss it.

I had a salt water aquarium when I was in college and even had a pet octopus for a while. They are intelligent and inquisitive. They can even taste things with the tip of the tentacles. I would set in a recliner near the aquarium when I studied and the octopus would get as close to me as it could and watch me. Sometimes I still miss that pet octopus, but I have many nice memories of it and the other sea creatures I have kept as pets.

Octopus, beautiful and intelligent

Many other fascinating creatures live in the seas. The largest animal that has ever lived is the blue whale. They are bigger than elephants. They are even larger than dinosaurs. Sometimes they can grow to over 100 feet long and weigh 150 tons. That is really huge. Think about it. One whale can weigh as much as 75 cars!

Blue Whale jumping out of the ocean
www.zmescience.com

God made the oceans and filled them with many wonderful creatures on the fifth day of Creation. We need to study and enjoy all of them.

The First Saturday

And God said, "Let the land produce living creatures according to their kinds: livestock, creatures that move along the ground, and wild animals, each according to its kind." And it was so. God made the wild animals according to their kinds, the livestock according to their kinds, and all the creatures that move along the ground according to their kinds. And God saw that it was good. Then God said, "Let us make man in our image, in our likeness, and let them rule over the fish of the sea and the birds of the air, over the livestock, over all the earth, and over all the creatures that move along the ground." So God created man in his own image, in the image of God he created him; male and female he created them. God blessed them and said to them, "Be fruitful and increase in number; fill the earth and subdue it. Rule over the fish of the sea and the birds of the air and over every living creature that moves on the ground." Then God said, "I give you every seed-bearing plant on the face of the whole earth and every tree that has fruit with seed in it. They will be yours for food. And to all the beasts of the earth and all the birds of the air and all the creatures that move on the ground-- everything that has the breath of life in it-- I give

every green plant for food." And it was so. God saw all that he had made, and it was very good. And there was evening, and there was morning-- the sixth day. (Gen 1:24-31)

This is the most important day of Creation. God made many new things on this day. He made all the land animals and told them to reproduce. He was pleased and said they were good. God also created people on this day. We were made in the image of God. He told us to rule over the animals and all the earth.

There is another important truth in this passage that many people miss. All animals and man were given only plants to eat for food. Yes, that means all animals were vegetarians. God forever changed this unusual practice changed hundreds of years later after Noah and the Flood. Today, both man and many animals eat other animals. Here is the command given after the Flood that remains in effect today.

Everything that lives and moves will be food for you. Just as I gave you the green plants, I now give you everything. (Geneses 9:3)

The First Sunday

Thus the heavens and the earth were completed in all their vast array. By the seventh day God had finished the work he had been doing; so on the seventh day he rested from all his work. And God blessed the seventh day and made it holy, because on it he rested from all the work of creating that he had done. (Gen 2:1-3)

God ended the creation process. It was finished. God rested on the seventh day…the first Sunday. This is repeated as part of the Ten Commandments.

For in six days the LORD made the heavens and the earth, the sea, and all that is in them, but he rested on the seventh day. Therefore the LORD blessed the Sabbath day and made it holy. (Exodus 20:11)

This is proof the six days of Creation were normal 24-hour days. Our worship on the seventh day is obviously a normal day as were the first six days of creation.

Several members of my family were charter members of Emmanuel Baptist Church when it was a small country church.   That was where I accepted Jesus Christ as my Savior and where as a lowly high school sophomore I felt God was calling me to ministry in the area of Creation and evolution and I have been doing this most of my life. The church had stopped growing and while I was in the Air Force it was moved to Weatherford and is now one of the major churches in town with an outreach around the world. I taught the college class while I was attending college and am proud my family had a small part in this important ministry. Here is a picture of the church today.

Emmanuel Baptist Church

     We must always attend church and worship our Creator on this special day. We are told in scripture to do this.

Let us not give up meeting together, as some are in the habit of doing, but let us encourage one another-- and all the more as you see the Day approaching.
(Hebrews 10:25)

Part 2: Examples of God's Creation

Through him all things were made; without him nothing was made that has been made. (John 1:3)

A plant that melts snow

Skunk Cabbage.
Photo by Norbert Smith

Birds and mammals are warm blooded. They can make their own heat to stay warm when it is cold. The temperature of other animals and most plants is determined by the environment. When it is warm they are warm, but on cold days their temperature drops. If the temperature drops below freezing the leaves of most plants freeze and die. That is why many trees lose their leaves in the winter.

The skunk cabbage is different. The tender leaves of the skunk cabbage cannot survive freezing. It is one of the first plants to come up in the spring. Just like warm blooded animals it produces its own heat. Even when the temperature is freezing it can remain at room temperature.

Skunk cabbage melting snow
ontariowildflowers.com

God truly made an amazing world of living things for us to learn about and enjoy. For this we must lift our hearts and our hands to praise Him.

You are worthy, our Lord and God, to receive glory and honor and power, for you created all things, and by your will they were created and have their being. (Revelations 4:11)

Things kangaroo rats have taught us

Cute and lovable Kangaroo rat.
Ss10e.blogspot.com

While walking in the Arizona desert long ago, I thought about kangaroo rats. They live on my farm in western Oklahoma, but they were abundant here with many burrows and fresh tracks. Kangaroo rats are beautiful little brown and white animals. They have taught us some important things.

Kangaroo rats make excellent pets. I kept several as a young boy growing up on a farm. This was before they were protected by law. To enjoy them now, you must watch wild ones or go to a zoo. They require little care and are fun to watch. They eat little and require absolutely no drinking water. Their not needing drinking water is unusual and important. It helps them survive in the desert where water is not available. There is more.

They can live on a diet of only dry seeds. They make all the water they need from the sugar in the things they eat. As we learn in science classes, when sugar is digested water and carbon dioxide are produced. This is the only water required by these delightful little creatures and they make it from the sugar in their diet. Their kidney concentrates water and as scientists studied how they do this, we learned many important things about how the human kidney works. Once again, asking questions of animals often provides us with knowledge and a better understanding of the world around us. What do you think about when you walk in the desert? I think about kangaroo rats and their Creator.

I praise you because I am fearfully and wonderfully made; your works are wonderful, I know that full well. (Psalms 139:14)

A plant that migrates

Frog surrounded by duckweed

The name "duckweed" is misleading because it is neither a bird nor an unwanted weed in a flowerbed. Duckweed is a small floating plant found in still water. Many of us have enjoyed it in our aquariums and you can buy them where tropical fish are sold. Duckweed seldom blooms, but when it does it has the smallest flower in the world. You must have a microscope to see the tiny flower.

Duckweed is unusual for another reason. It is the only plant that migrates. When one thinks of migration, animals such as birds, fish and perhaps a few mammals come to mind. Many of our song birds and waterfowl fly south each winter and return in the spring. Elk migrate above the timber line in summer and move down the mountains during winter. In centuries past, the American buffalo migrated during the warmer months from northern Mexico to southern Canada in herds so large it sometimes took several days for them to go past any given location. Even monarch butterflies migrate thousands of miles each year. In the summer they live throughout the United States and southern Canada. As winter approaches they migrate to Mexico. It is still not understood how they make this remarkable journey each year.

Like many birds that fly south for the winter, duckweed also migrates and their migration is fascinating and helps them survive where the winters are cold. Duckweeds are found over much of the world including where winters are very cold. They survive the cold winter months by producing tiny buds that sink to the bottom of the pond or migrate. The buds remain on the bottom in the cold dark water below the ice. In spring as the ice melts and the light increases, oxygen is produced and they migrate back to the surface of the pond for another year. These tiny plants are important to the pond by providing shade which lowers the water temperature. The cooler water increases the oxygen in the water. Oxygen is needed by fish and other creatures living in the water. They also reduce pond water loss through evaporation by covering the surface.

If God cares for and protects this tiny plant, think how much more He cares for and will protect you. God loves children as the scripture below clearly shows.

Jesus said, "Let the little children come to me, and do not hinder them, for the kingdom of heaven belongs to such as these." (Matthew 19:14)

Alligators have leaking hearts

Animals with backbones are called vertebrates. They include fish, amphibians, reptiles, birds and mammals. The hearts of each of these animals are made differently. Fish have a simple heart with only two chambers. Amphibians have a more completed heart with three chambers. Alligators, birds and mammals have a four chambered heart. The right side pumps blood to the lungs so it can pick up oxygen and release carbon dioxide. The left side of the heart pumps blood to the rest of the body. The body uses oxygen and produces carbon dioxide. Blood returns to the heart and is pumped again and again.

Alligators and crocodiles are found around water and often dive underwater for food or to sleep. God made their heart special so they can dive longer. Some can stay under water for more than two hours. They are also the largest of all reptiles and a crocodile caught in the Philippines was over 21 feet long and weighed 2,370 pounds. When they are at the surface breathing air their heart works like the heart of birds and mammals. The left side pumps blood to the lungs to get oxygen and release carbon dioxide. When they are underwater and unable to breathe, an opening in the heart opens. This opening allows blood to bypass the lungs and helps them stay underwater longer.

God gave all animals and plants exactly what they needed to survive. God also gives you what you need to survive and be happy. We must always remember to thank and praise God for all he does.

I will greatly praise the LORD with my mouth. (Psalms 109:30)

A beetle with a powerful weapon

Bombardier Beetle.
Wikipedia.org

All animals have the ability to escape from enemies. Many animals run and hide, but some have a weapon and can hurt the creature that attempts to attack them. The Bombardier beetle is perhaps the best example. There are over five hundred kinds of bombardier beetles. They all have a powerful chemical weapon. They can accurately fire a boiling hot foul-smelling liquid at an enemy. The expulsion also makes a loud popping sound. The sound provides additional protection. Bombardier beetles produce and store two powerful chemicals which collect in a reservoir until they are needed. These toxic chemicals are not found in any other living creature. When threatened, the two chemicals are squirted into a chamber where they are mixed with an enzyme. The resulting mixture boils and is directed to the attacking predator. The damage caused can be fatal to attacking insects and small creatures and is even very painful to human skin. Such a complex weapon is proof of the wisdom of its Creator. The Psalmist said it best:

The fool says in his heart, "There is no God."

(Psalms 14:1a)

Armadillos are unusual creatures

Armadillo digging for worms
Photo by Norbert Smith

Armadillos are one of the strangest animals that God created. The word "armadillo" means little armored one in Spanish. The Aztec people of Mexico called them "turtle-rabbits". Armadillos live in Oklahoma, but have not been here very long. They crossed the Red River between Oklahoma and Texas in the late 1950's and now they live as far north as Nebraska and Illinois. They are expanding their range in North America because they have no natural predators here.   They eat mostly earthworms, but surprisingly they do not hibernate. They are out all winter looking for worms. In the northern parts of their range, many have lost the end of their tail and tips of the ears frozen off.   Much like a turtle, their hard leathery shell protects them from most predators. Although their armor-like skin is their most important defense, they can also escape predators by running away and often run into thorny bushes where their armor protects them.

They are one of the few animals that cannot swim without preparing first. Norbert Smith has tested this by tossing them into a creek and they sink to the bottom and walk out the other side. They can hold their breath for six minutes. They can swim slowly, but must first swallow air and inflate their stomach to twice the normal size in order to float and remain above the water.

Perhaps their most unusual trait is the mother armadillos always have four identical babies or quadruplets. Because of this, they are often used in genetic studies. They are also the only animal, besides humans, to get leprosy and are used to study this dreaded disease. Armadillos are interesting animals and can teach us many things if we will study them. Sadly many armadillos are killed by cars because they jump up as the car passes over them on the road.

For everything God created is good. (1 Tim 4:4)

Why this wasp taps on trees

Ichneumon wasp
Photo by Gene Hicks

This beautiful wasp has many features that display the Creator's wisdom. These insects look more like a fly than wasp. There are 60,000 kinds of ichneumon wasps world wide and 3,000 in North America. It is the most abundant wasp. They are unusual in that they are not found near the equator because their young feed on insects that feed on hardwood trees not found in jungles. Unlike other wasps they do not sting. Both males and females are often seen tapping on trees with their antennae. Males do this to find females. Females do this to find wood boring larvae in which to deposit her eggs. No one knows how she can force the soft flexible egg laying organ into hard wood. Scientists have recently found metal manganese and zinc at the tip of the organ. These metals are not found in any other animal. It is unknown how she can locate the exact location and size of the larva inside the tree trunk just by tapping on the wood.

There is more. The adult insect is faced with the seemingly impossible problem of escaping from deep inside the tree. Again the high metal concentrations are also found in the mouth parts of the adult. This is how she chews herself out of the wood in order to find a male and repeat the circle of life. These are truly amazing creatures and show evidence of being designed by an all knowing Creator.

Through him all things were made; without him nothing was made that has been made. (John 1:3)

Giraffes have a big heart

So God created the great creatures of the sea and every living and moving thing. (Genesis 1:21)

Giraffes are very tall.
View Source: flickr.com

When people see a giraffe for the first time, they are amazed at their height. They are the tallest of any animal and can get up to 20 feet tall or nearly as tall as 4 people. When I saw my first giraffe, I asked my beloved Grandfather, "How does its heart get blood up to its head?" That question has been on my mind for many years and now I understand how it does this amazing thing.

To get blood up to the brain they must pump blood up nearly 12 feet. To do this their hearts are really large and their blood pressure is higher than it is for any other animal.

Once again the more we learn about how complicated living things are…the stronger is the evidence they were designed by an all knowing Creator.

For by him were all things created, that are in heaven, and that are in earth, visible and invisible…all things were created by him. (Colossians 1 1:16)

View Source: pinterest.com

Ants work really hard

Ants are creatures of little strength, yet they store up their food in the summer. Proverbs 30:25

It seems ants are always busy. I enjoyed watching them as a boy and even today I like watching them work. They have a queen ant, but she does not tell them what to do. God programmed each worker ant to work hard to find and bring food to the young ants in the den. They often gather food far from their home and scientists have recently discovered how they find their way back. They have tiny magnets in their antennae that help them find their way. It is much more complicated than a simple magnetic compass. It is a complex system much like the modern GPS systems that use satellites that uses orbiting

4,000 pound satellites and cost millions of dollars.

Go to the ant, you sluggard; consider its ways and be wise! It has no commander, no overseer or ruler, yet it stores its provisions in summer and gathers its food at harvest. (Proverbs 6:6-8)

Buffalo were designed to die

Magnificent American Buffalo
Photo by Norbert Smith

For thousands of years Native Americans depended on the buffalo for survival. Every part of the animal was used and nothing was wasted. The meat provided food and the hides were used to make tents and clothing. The bones were used to make tools. Their disappearance marked the end of a way of life. People all over the world were impressed that Native Americans could kill these huge beasts using only a bow and arrow. Large bulls can weigh over 2,000 pounds and are very mean.

The American Buffalo once lived from northwest Canada, south across the grasslands of the western United States. Some migrated herds migrated south into central Mexico in winter. Some experts thought there were as many as 70 billion buffalo and perhaps more. Some of the old buffalo trails can still be seen in Mississippi and Tennessee.

The twice-a-year migrations were impressive. Their approach sounded like thunder and the ground shook as in an earthquake. Some of the larger herds took days to pass. The larger herds emptied small lakes by drinking all the water. There has not been anything like it before or since.

There were several ways bison were killed by Native Americans before horses were available. They could be shot from a blind or other hiding place using a bow and arrow. Sometimes Native Americans wore wolf skins and got close enough to shoot a few bison with bow and arrow. This method was risky as bulls sometimes charged at the intruders, but it provided food, skins and tools. Here is a painting from the Gilcrease Museum in Tulsa, Oklahoma.

Sneaking up on a herd of buffalo.

The introduction of horses changed the lives of native people. Horses provided rapid transportation and more effective methods of harvesting bison. The simplest method was running down a single bison by horse. It sometimes took as many as five or more, fresh horses to fatigue a healthy bison.

The method seen most often on TV involves a rider on a horse shooting a running bison with a bow and arrow. This problem has an unlikely solution. The evidence suggests the American buffalo was uniquely designed by the Creator to feed a people.

George Catlin, *Buffalo Chase*, 1845

All other mammals have two separate lung cavities. One side of the human chest, for example can be penetrated, collapsing that lung, but the remaining lung continues to support life. God created the buffalo with both lungs in one cavity. The problem for the Native bow hunter is solved. An arrow anywhere in the chest collapses both lungs and provided food. Truly God created buffalo easy to kill for Native Americans.

The heavens declare the glory of God; the skies proclaim the work of his hands. Day after day they pour forth speech; night after night they display knowledge. There is no speech or language where their voice is not heard. (Ps 19:1-3)

Ducks have cold feet

Mallard duck.
View source: publicdomainpictures.net

One of the reasons I like zoology is because of the special features God gave some animals to survive in harsh environments. Ducks provide an excellent example. Certainly their annual migration over thousands of miles is impressive, but I am more impressed with their feet. Certainly the webbing between the toes provides excellent propulsion for swimming, but there is more. Ducks return to their summer nesting areas when it is still cold at night and the water is near freezing. They spend most of their time swimming in icy cold water. A duck's feet are large, yet they lose very little body heat. How is this possible? If our hands or feet are in ice water for even a short time we begin to feel cold all over because of the heat loss. Certainly a duck's body is covered with warm and water repellant feathers and down. But their feet are in direct contact with ice cold water. Why do the feet not lose their body heat to the cold water?

The answer is amazing and again shows proof of an all wise Creator.   The warm arterial blood from the heart flows near the cold blood returning from the foot. The blood vessels break up into small vessels that intertwine with each other. The blood is flowing in opposite directions and the warm arterial blood is cooled by the cold blood from the foot. The cold blood from the feet is warmed from the arterial blood and little heat is lost or wasted.

The evidence is conclusion and obvious. Ducks were designed to survive swimming in ice cold water while maintaining their warm body temperature.   The next time you see ducks or geese swimming in cold water remember their Creator. It was He who designed them so their feet could be cold without losing their body heat. What a truly amazing God we worship. We must always remember to praise Him for the wonderful Creation we enjoy.

All flesh is not the same: Men have one kind of flesh, animals have another, birds another and fish another.
(1 Corinthians 15:39)

Hummingbirds hibernate on cold nights

Humming bird in flight.
View Source: Hummingbird by Ruthie True from Pinterest

Hummingbirds have always impressed my family. My grandmother planted honeysuckles near the house to attract them. I remember watching them as a kid and was sometimes able to get close enough to hear the distinctive hum of their rapid wing beats. I was impressed how they could hover near a flower and even fly backwards after harvesting the nectar. Later as an adult, I put up a hummingbird feeder and still enjoy watching them. Their small size and helicopter-like flight have always fascinated me, especially after owning and flying my own airplane.

Let me share some of the things that make hummingbirds unique.

First, they are indeed the smallest bird and are the smallest of all warm blooded or endothermic animals. The shrew is next and is not much bigger. Yes, I have even found a couple of their little nests. They line them with soft spider webs to protect their delicate pea sized eggs. As one might expect due to their small size, hummingbirds in flight have the highest rate of metabolism of all animals except a few flying insects. Their heart rate can reach 1,260 beats per minutes. Even at rest, the have an extremely high rate of metabolism and are always but a few hours from starving to death. It is uncanny that the tiny ruby-throated hummingbird is able to migrate 800 KM (500 miles) across the Gulf of Mexico twice a year. There is another more frequent problem.

They barely have the ability to store enough energy to stay alive on warm nights. In order to survive long cold nights they must reduce their rate of metabolism even farther especially when food is in short supply. They do this by lowering their body temperature. In other words they go into nightly hibernation more accurately known as torpor. In this state, their heart rate slows to 50 to 100 beats per minute and their body temperature drops to near the ambient temperature as is the case for hibernating mammals. While this solves the problem of starvation at night it creates another problem for females incubating eggs. The time required for the incubation of most birds is well known and fixed. With hummingbirds it is highly variable. The reason is simple. If it is cold and the bird must go into nightly torpor it takes longer for the eggs to hatch. Incubation typically takes from 14 to as long as 23 days depending on the weather.

Once again this presents a plethora of problems for the evolutionist. Since it is so effective for hummingbirds, why has it not evolved for other endothermic animals? It requires many special enzymes and other compounds for neurons to operate over the wide temperature ranges. Once again the behavior, physiology and biochemistry must all be present at the same time for them to accomplish these remarkable feats. It is impossible to conceive of this complex response developing in small steps caused by mutations or genetic errors. Once again evolution has failed to explain the facts of science.

Are not two sparrows sold for a penny? Yet not one of them will fall to the ground apart from the will of your Father. (Matthew 10:29)

Opossums are good actors

If Oscar awards for acting were given to animals, opossums would win every year.

Virginia opossum playing dead.
Photo by Norbert Smith

Opossums are found throughout much of central and North America and range as far north as northern Minnesota. They are best known for playing dead and I studied them on my farm with Dr. Geir Gabrielsen from Norway. We met when I lectured at Oxford University in England and have stayed in touch over the years. We used my pet dog to elicit the response. While playing dead they were unresponsive and their heart rate dropped to less than half the normal rate. Breathing also slowed and they turned blue from the lack of oxygen. There was no normal blink response when the eye was touched. This is why some scientists thought they were unconscious. Without farther disturbance the pre-stimulus values slowly returned. In spite of my scolding, the dog returned and even without touching the opossum its heart slowed again. It was obvious the opossum fully conscious because as the dog returned its heart would again slow.

This study took an unexpected twist. My personal motivation to study animals is to know more about the animals. I do not study animals to relieve human suffering, cure disease or prolong life. I study them because I want to know more animals.

It was Geir who made the connection to human illness. He was working at a medical research laboratory that was studying Sudden Infant Death Syndrome or SIDS. He returned to the laboratory excited about our opossum data. We had taken several photographs during the experiments and he showed a pediatrician the photo above of an opossum playing dead. Geir was ecstatic because he and I now had a deeper understanding than anyone else in the world about what happens when an opossum plays dead. Together we had gotten a glimpse of yet another mountain not previously seen. We were also developing a clear picture of the passive fear response which we both had been studying for years. The pediatrician was ecstatic for a different reason. Upon seeing the photo and hearing Geir describe death feigning in opossums, the doctor saw a piece to a very different puzzle.

Sudden Infant Death Syndrome or SIDS is the largest cause of death for children from 6 weeks to 2 years of age. For no apparent reason they stop breathing and die. Unlike other infant deaths there is no obvious cause. Although the cause of death is lack of oxygen from failure to breathe, no one knows what triggers the fatal episode. Geir and the physician performed a simple experiment. They approached cribs where babies at risk for SIDS were sleeping. Geir clapped his hands loudly one time. Most of the babies suddenly held their breath, turned blue, and might have died had not the physician been nearby to resuscitate them. It appears a baby at risk for SIDS is frightened to death by a loud noise. This was an important piece to the puzzle of SIDS – a piece discovered by accident one night with an opossum and a disobedient dog.

Our discovery has saved the lives of thousands of babies. For example if a nurse working with babies at risk for SIDS tips over her crash cart, she is told to check the babies before picking up her mess. Of course, there is still something else wrong with the child because fear should not cause an infant to stop breathing until he/she dies. There is a much more related to SIDS and other scientists continue to fit those pieces together. It is satisfying to know that my own research played a small part in understanding this tragic illness.

This illustrates how science advances. Many important discoveries were made by accident while working on something else. A scientist is simply following his/her curiosity about the world we share. No one knows where a discovery might lead. And it is this unknown that makes scientific research an exciting and rewarding pursuit.

Reference

Geir W. Gabrielsen and E. Norbert Smith, 1985 Physiological responses associated with feigned death in the American opossum
Acta Physiol Scand (April) 123(4):393-8.

This wasp is a surgeon

© Z. Huang

Cicada Killer Wasp with prey
Photo by Zachary Huang, www.cyberbee.net

Introduction

The beautiful yellow and black cicada killer, *Sphecius speciosus* is the largest wasp in North America and can reach a length of 5.08 cm (2 inches). Although they are huge and appear menacing, they are normally not aggressive. Adults feed on the nectar from a variety of flowers. Cicada killers are found from the Midwestern United States south across Mexico and into Central America. They are ground burrowing solitary wasps classified in the family Crabronidae. Males do not sting, but are often seen in groups challenging one another for a female. It is not unusual to see two or three males flying erratically locked in midair combat. Females often share their large burrow with other females. Each burrow consists of several chambers or rooms with each one containing from 1 to 3 cicadas.

As a boy growing up in western Oklahoma, I often watched these amazing creatures as they captured, immobilized and finally flew off with the huge paralyzed cicada. The process often took more than an hour to complete. Cicadas are known locally in Oklahoma as "locusts" and they are three times larger than the wasps that capture them. Many years later, as a zoologist, I came to understand and appreciate some of the details of this complex and highly orchestrated feat. This is another example for which evolutionists have no rational explanation. It shouts DESIGN and where there is design, there must be a Designer. Sadly, any discussions of the failure of evolution are strictly banned in the university classrooms. Our students are being deceived by the professors that are charged with teaching them to think.

Discussion

Large female cicada killer wasps are commonly seen skimming across lawns looking for nest sites or searching for cicadas in trees and shrubs. Their reproduction process is complex and involves specialized anatomical structures as well as several intricate behavioral steps. First, the female digs an extensive underground burrow 25 to 50 cm (10-20 inches) deep and 1.5 cm (0.59 inches) wide. In order to dig the burrow, the female dislodges the soil with her jaws and using her hind legs, pushes the loose soil behind her as she backs out of the borrow. Her hind legs are equipped with special spines that aid in the process of pushing the dirt out of the hole. The burrow is sometimes shared with other females and has several individual compartments or rooms where the cicadas will be placed later and where the young will develop and grow.

After the burrow is complete, she must find and sedate the large and elusive cicada. Upon finding a cicada she first systematically stings the nerve ganglia that control the flight muscles, rapidly subduing the huge struggling insect. When the wings have stopped thrashing about she then systematically stings the ganglia associated with movement of each of the six legs. Once the prey is effectively paralyzed, she crawls up the nearest tree trunk or fence post dragging the huge cicada beneath her. She then flies toward the previously dug burrow clutching the large prey tightly with their legs. The weight is too much for her to maintain altitude and she gradually sinks to the ground. Upon landing, she again drags the heavy prey up a nearby post or tree trunk and repeats the process until she finally arrives at the burrow. She then places it in one of the previously built chambers drags and lays a single egg in it. The entire process often takes an hour or more. Like queen bees, she instinctively knows which egg will develop into a male or female. She often puts two or three cicadas together for depositing eggs to produce females, but puts a single cicada in a chamber to deposit the egg that produces a male. A single borrow may contain 10 or more separate chambers or rooms. The eggs hatch in 1 or 2 days and the larva complete their development in about 2 weeks. The larva instinctively avoids eating the vital organs until it is nearly mature. Finally, it eats the vital organs, killing the cicada and quickly pupates. Pupation lasts about a month and there is only one generation per year.

Conclusion and implications

Even as a young boy, I was impressed at the instinctive knowledge the adult cicada wasp had of the cicada's nervous system and how to sting it without killing the prey. I was also amazed with their ability to find the burrow that had been dug hours before capturing the prey. Each of the components of this complex behavior must be complete and genetically passed on to the next generation. Such detailed behavior patterns could not be discovered anew with each hatchling. Nor could they have developed by random mutational errors over time. Evolutionists have no rational explanation for the complex behavior.

Stop and consider the complexity and overwhelming evidence of design clearly displayed for all to see. It is utterly impossible for a female wasp to accidentally find each of the nerve ganglia that control movement in the giant cicada or accidently know how much venom to inject. Too much would kill the prey and her young would starve...too little and it would fight the young again causing it to starve. No matter how much time was involved such complex and highly orchestrated behavior could not have happened by random chance alone. The entire process must also be genetic and passed on to the young. The complexity displayed here shouts DESIGN as few other things in nature. Where there is design there must be a Designer. It should be obvious to all people with an open mind...that Designer is none other than the Creator of all things described in Genesis.

It boggles my mind that some actually believe all this happened by accident in many tiny steps by trial and error of vast periods of time. Remember, each step MUST be advantagenous or it will be eliminated by natural selection. Each step must also be genetically passed on to the next generation or it will be forever lost. If you believe all that developed by accident, might I sell you some beach front property in Arizona? The evidence of design is overwhelming in this example alone. What am I missing? How can evolutionists not see the evidence? The unmistakable evidence of God is all around us. Even as a boy and especially now as an experienced scientist, I see the evidence of God as Creator is overwhelming. I am not alone in this view as tens of thousands of former evolutionists have abandoned evolution dogma like rats from a sinking ship. (See Jerry Bergman and other references in the reference section below.)

I find it strange with all we know today about the complex structure of all living cells and their amazing behavior that anyone can still cling to the belief that all this just happened without direction in small inheritable steps by random mistakes in the genetic code. Scripture has the answer to this dilemma. *For this reason God sends them a powerful delusion so that they will believe the lie.* (II Th. 2:11, NIV) That delusion is evolution. Truly, as God's Word declares only the fool fails to see the hand of God in nature.

The fool says in his heart, "There is no God." (Ps 53:1a, NIV)

Those failing to see the Creator are without excuse.

The heavens declare the glory of God; the skies proclaim the work of his hands. Day after day they pour forth speech; night after night they display knowledge. There is no speech or language where their voice is not heard. (Ps 19:1-3, NIV)

References

Bergman, J. 2011a. Slaughter of the Dissidents. Leafcutter Press.

Bergman, J., 2011b. The Dark Side of Charles Darwin. Master Books.

Dambach CA, Good E. 1943. Life history and habits of the cicada killer (Sphecius speciosus) in Ohio. Ohio Journal of Science 43: 32-41.

Evans, H. and Kevin O'Neill. 2007. The Sand Wasps: Natural History and Behavior. Harvard University Press pp. 37-43. ISBN 978-0-674-02462-5.

Holliday CW. (2009). Prof. Chuck Holliday's Cicada Killer Page. Lafayette College. http://ww2.lafayette.edu/~hollidac/cicadakillerhome.html (20 July 2009).

http://entomology.ifas.ufl.edu/creatures/beneficial/cicada_killers.htm

Smith, E. N., 2010. Creation or Evolution? Consider the Evidence before deciding.

Smith, E. N., 2011a. Evolution has Failed.

Smith, E. N. (editor), 2011b. Sacred Cows in Science, no objectivity allowed.

Smith, E. N. and S. K. Kern, 2012. Creation in Six Days.

Smith, E. N., 2012. Evolution in Disarray. In press.

Wikipedia: Sphecius speciosus

Dancing bees

Happy dancing

Bees are important to farmers and they provide honey for us to eat. Bees pollinate crops so they can produce seeds. Without bees many crops would die. Many bees in the United States have died from a terrible disease. Some places have lost all of the wild bees but there are now some bees that do not get the disease.

People have collected honey from bees and raised them for thousands of years. I have kept bees for many years and enjoy the fresh honey. Here is a photograph of one of my beehives.

Sometimes more than 50,000 bees live in a beehive. Bees need nectar and pollen from flowers for food. They also need it to feed their young.

A typical beehive
Photo by your author

Flowers do not last long. Many flowers bloom and die quickly. How do bees living inside a beehive know where new flowers are blooming? Bees have an unusual way of telling the other bees where flowers are found. A bee that finds a new patch of flowers does a dance on the honeycomb inside the beehive. The dance tells the other bees which direction the flowers are located from the beehive and how far away they are. She also gives tiny samples of honey to the other bees so they will know when they have found the new flowers.

Honey is mentioned many times in the Bible. Samson was an important judge in the Bible. He was a very strong man. He was attacked by a lion and killed it with his bare hands. A few days later a swarm of bees moved into the lion skin. He made up a riddle about the lion and bees. Here is his riddle from the book of Judges in the Bible. ***Out of the eater, something to eat; out of the strong, something sweet.***

No one could figure out his riddle because they did not know about the lion and the bees. The Bible has humor and even riddles in it. We need to read and study it every day. We can learn many important things from the Bible.

Be diligent to present yourself approved to God, a worker who does not need to be ashamed, rightly dividing the word of truth. (2 Timothy 2:15)

Bumblebees hum while they work

Bumblebee looking for food.

Many people hum or sing while they work. Do you? We enjoy the beautiful sound birds make when they sing. We also like the sound of a cat purring. Humming is important for bumblebees for another reason. It helps them find food.

Bumblebees do something called "buzz pollination." Some flowers have their pollen tightly locked away, but still need to be pollinated. For some of these flowers a loud sound of a certain frequency causes the flower to release the pollen to anyone singing the right note.

The bumblebee uses its flight muscles to make a loud buzzing sound, without noticeable wing movement. The sound releases a cloud of pollen easily seen by anyone watching. The buzz is distinct in pitch and noticeably louder than the normal buzzing sound associated with bee flight.

Here is another example. Tomato pollen is locked inside the flower and is released only when a sound is present. Honeybees work in silence and thus no pollen is released and they collect scattered pollen left behind by the most recent buzz pollinator. Bumblebees hum while they work at EXACTLY the proper frequency to release the tomato pollen and a small cloud of yellow pollen appears. These profound observations demand a Designer and we can know the Designer of flowers and bumblebees.

You are worthy, our Lord and God, to receive glory and honor and power, for you created all things, and by your will they were created and have their being. (Revelations 4:11)

Ants use GPS to navigate

Ants are creatures of little strength, yet they store up their food in the summer. Proverbs 30:25

      The network of Global Positioning Satellites (GPS) has forever changed the way we navigate. Designed for the military, it is now widely used by anyone with a GPS receiver. GPS navigation capability is commonly used in automobiles to help drivers travel cross-country without the need for a roadmap. Delivery trucks use them to find customers for package pick-up and delivery. Small portable models are available to help hikers avoid getting lost in remote wilderness areas.

Research with ants has revealed that they have been using a GPS-like navigation system long before man invented it. Tiny magnets recently found in the antennae of ants help them find their way. The magnetic antenna system is thought to work with iron particles in the soil to create a kind of highly sophisticated natural GPS system. The system appears to be built out of dirt.

Human global positioning systems rely on power-consuming receivers that pick up information from clunky, orbiting 4,000-pound satellites costing millions of dollars to develop and launch into orbit. In sharp contrast, the ant GPS system weighs next to nothing, requires little energy to operate and is earth-friendly to the maximum. The study was conducted by researcher Jandira Ferreira de Oliveira of the Technical University of Munich and the Brazilian Center for Physics Research. "The ants we studied dwell in tropical soils that are full of very fine-grained iron minerals, so there is plenty of material available," said Dr. Oliveira. "The incorporation of minerals probably starts as soon as ants start getting in touch with soil," she added, explaining that her team found ultra-fine-grained crystals of magnetic magnetite, maghemite, hematite, goethite, and aluminum silicates in ant antennae. These particles could make a "biological compass needle" that drives the ant GPS.

For the study, published in the ***Journal of the Royal Society Interface,*** Dr. Oliveira and her colleagues collected worker ants from the species *Pachycondyla marginata* in Sao Paulo, Brazil. Prior studies found these ants tend to migrate at an orientation of 13 degrees relative to Earth's geomagnetic north-south axis, and that the ant's strongest magnetic signal comes from its antennae. High-powered microscopes and chemical analysis revealed the presence of the dirt-acquired magnetic particles in the antennae, intriguingly next to a body part called Johnston's organ that may also be part of the ant's GPS. Dr. Oliveira explained how the system evidently functions. Our planet is magnetized, likely due to rotational forces of liquid iron in earth's core. Although the resulting magnetic field is only one-twenty thousandth as strong as a refrigerator magnet, ants appear to "perceive the geomagnetic information through a magnetic sensor (the dirt particles), transduce it in a signal to the nervous system and then to the brain," she added.

Other ant species use different systems. Desert ants, for example, appear to possess special features in their eyes that can detect skylight polarization, which they then use to find their way around their sandy habitat. Magnetic particles, however, have been detected in fish, birds, butterflies, flies, bees, bats, mole rats, newts, sea turtles and spiny lobsters, suggesting these animals may find their way like the Brazilian ants do.

An almost identical GPS system might operate in rainbow trout and homing pigeons. Dr. Gerta Fleissner of the University of Frankfurt and her colleagues discovered maghemite and magnetite in the skin lining the upper beaks of these birds famed for their directional skills. Dr. Fleissner believes this "pigeon-type receptor system might turn out to be a universal feature of all birds." The University of Oxford's Robert Srygley, one of the world's leading insect experts, told *Discovery News* that the new study "is a major advance toward finding the magnetic compass in this nomadic ant." Once again it seems God in His infinite wisdom has given even His smallest creatures everything they need to find food and return safely home.

Questions for evolutionists

How did ants develop the complex ability to find their way through incorporation of metallic elements into their bodies? Evolution postulates that early ants would have developed from even earlier creatures without the ability to do so. Evolutionists would hypothesize that ants developed the ability through random mutation. Focus on the first ant with a random genetic mutation that allowed it to successfully incorporate the magnetic elements of ingested soil into his body. How would it acquire the knowledge to know what to do with this new information? How would it develop the corresponding behavior to respond to that information? It is impossible to believe that this ant mutant was simultaneously equipped by evolution with (1) the ability to incorporate magnetic particles, (2) a full set of information with what the new data meant to its survival, and (3) a unique knowledge regarding how to respond to the data in a way that conferred a survival advantage in feeding and breeding. This set of coincidences occurring in nature stretches credulity to the breaking point.

Once again we find Scripture eloquently says at the end of Creation week that:

God saw all that he had made, and it was very good. And there was evening, and there was morning -- the sixth day. Genesis 1:31

References

de Oliveira, J.F., 2009. *Magneto-ants pump iron*, **The Scientist**.
Viegas, J., 2009. *Magnets in Ant Antennae Work as Internal GPS*, **Discovery News**.
http://dsc.discovery.com/news/2009/05/20/ant-magnet-gps.html

You can know Jesus

"Whoever believes in Him shall not perish but have eternal life."
(Special courtesy to Reese Camp-Colonial Hill Baptist Church.)

At that time the disciples came to Jesus and asked, "Who is the greatest in the kingdom of heaven?" He called a little child and had him stand among them. And he said: "I tell you the truth, unless you change and become like little children, you will never enter the kingdom of heaven. Therefore, whoever humbles himself like this child is the greatest in the kingdom of heaven. "And whoever welcomes a little child like this in my name welcomes me. (Matt 18:1-5)

Here is some really exciting news. ALL people can personally know the God of Creation. If you want to do this, simply pray the following prayer…and mean it. You can trust Him. He will bless you and keep you safe.

Dear Lord Jesus,

I know that I am a sinner, and I ask for your forgiveness. I believe you died for my sins and rose from the dead. I turn from my sins and invite you to come into my heart and life. I want to trust and follow you as my Lord and Savior. Amen.

God loved you so much that He gave his son to die for you so you can go to heaven. Jesus lived a perfect life without sin and died a terrible death on a cross. He was buried in a tomb for three days and came back to life.

For God so loved the world that he gave his one and only Son, that whoever believes in him shall not perish but have eternal life. (John 3:16)

If you pray this prayer you can become a Christian and know the God of Creation and His Son, Jesus.

Dear Lord Jesus, I know that I am a sinner, and I ask for your forgiveness. I believe you died for my sins and rose from the dead. I turn from my sins and invite you to come into my heart and life. I want to trust and follow you as my Lord and Savior. I pray this in your name. Amen.

If you prayed that prayer and meant it you are now a Christian. To learn more about the Lord you need to read your Bible every day and join a church and make new Christian friends.

Go to the ant, you sluggard; consider its ways and be wise! It has no commander, no overseer or ruler, yet it stores its provisions in summer and gathers its food at harvest. (Prov 6:6-8, NIV)

Consider the ravens: They do not sow or reap, they have no storeroom or barn; yet God feeds them. And how much more valuable you are than birds! Who of you by worrying can add a single hour to his life? Since you cannot do this very little thing, why do you worry about the rest? "Consider how the lilies grow. They do not labor or spin. Yet I tell you, not even Solomon in all his splendor was dressed like one of these. If that is how God clothes the grass of the field, which is here today, and tomorrow is thrown into the fire, how much more will he clothe you, O you of little faith! And do not set your heart on what you will eat or drink; do not worry about it. (Luke 12:24-29, NIV)

About the author

Doc Gator with a young Cayman.

Norbert Smith, or "Doc Gator" as his university students called him, has spent most of his life studying alligators and other wild animals. He has written over 100 scientific articles, books and magazine articles. He was the first scientist to study the heart rates of animals when they are frightened and hiding. He discovered their hearts slowed when they have a safe place to hide from danger. After his honorable discharge from the Air Force, he worked in electronics for three years and wrote articles for electronic magazines such as *Popular Electronics* and *Electrical Design News*. He also wrote stories about animals for the children's magazines *Highlights* and *Ranger Rick*. He has caught, studied and released over 200 wild alligators in south Texas. One of the alligators weighed over 750 pounds. He has also studied captive animals in zoos. His alligator research was featured in the BBC TV documentary, *A smile for the Crocodile*. He was invited as keynote speaker at an international radio telemetry conference at Oxford University in England and took three of his research students with him. He enjoys studying the animals God created.

Dr. Smith is retired and lives on the family farm of his childhood where he enjoys gardening his two acre vegetable garden. He enjoys retirement very much spends most of his time writing books, having written and published 19 books. Seven are children's books. You can find his books online or have your bookstore order them. Check out his website, *www.GodofCreation.com.* You may email him at *Docgater@aol.com* or snail mail at 24340 East 1080 Road, Weatherford, OK 73096.

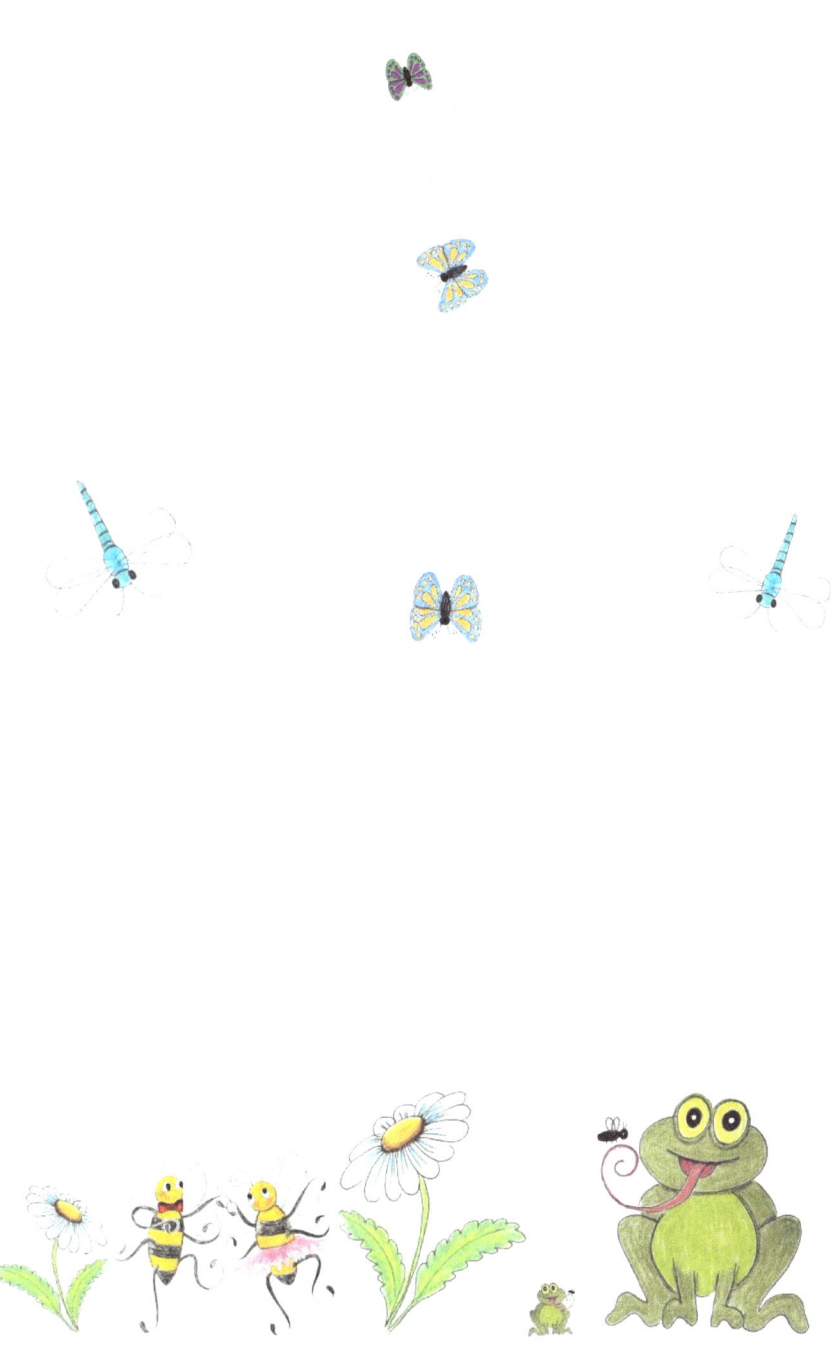

About the Illustrator

Nathalie Kelley, Illustrator

Nathalie Kelley is a full time, award winning artist. She was born into an artistic Texas family where both parents and grandparents are professional artists and were an inspiration for Nathalie's earliest memories to be about art.

She is a versatile artist in her subject and media. Her training in her youth was in oils and primarily landscapes. In the 1990's she was captivated by the beauty and challenges of watercolors. Her art includes realism in watercolor, worship art in acrylic on canvas, and hand painted silk worship flags, streamers and scarves. Her art work ranges from the small, tiny pieces to wall-sized murals.

Her internationally collected art work graces numerous businesses, homes, offices, and churches. Her commissioned works include various murals for private and corporate collections, portraits, silk flags for worship celebrations and the illustrations for a children's book "The Little Lamp" and "It Happened One Day in Texas".

Nathalie and her husband, Joe, live in Snyder, TX. For further information, contact: info@nathaliekelleyart.com or Nathalie Kelley, 3798 Dalton Dr., Snyder, TX 79549, ph. #325-207-6474. www.nathaliekelleyart.com .